A Ladybird

Ruby Tuesday Books

Ruth Owen

Published in 2025 by Ruby Tuesday Books Ltd.

Copyright © 2025 Ruby Tuesday Books Ltd.

All rights reserved. No part of this publication may be reproduced in whole or in part, stored in any retrieval system, or transmitted in any form or by any means, electronic, mechanical, photocopying, recording, or otherwise, without written permission from the publisher.

Editor: Mark J. Sachner
Design: Tammy West
Production: John Lingham

Photo Credits:
Alamy: 16 (Jurgen Kottmann); Nature Picture Library: 7 (Stephen Dalton), 13 (Nature Production); Science Photo Library: 20–21 (Claude Nuridsany & Marie Perrenou); Shutterstock: Cover (Fariya Graphics), 4 (nieriss), 5 (Mickey Why), 6 (Cornel Constantin), 8 (JulieK2), 10 (irin-k), 11 (Sinan Guzel), 12 (Susanne Photography & Rina 991), 14 (Tabish Hassan Khan), 15 (Distracted By Bugs), 17 (Radzas2008), 18 (Szabadi Jeno Tibor), 19 (Neil Bromhall), 22 (Tom Deer, irin-k, mehmetkrc, & chinahbzyg), 23 (Smeerjewegproducties, aaltair, & Protasov AN), 24 (mehmetkrc); Superstock: 9 (Matt Cole).

ISBN 978-1-78856-439-7

Printed in Poland by L&C Printing Group

www.rubytuesdaybooks.com

CONTENTS

Hello, Little Ladybird! 4

Glossary 22

Index 24

Hello, Little Ladybird!

It is morning in a sunny garden.

Bug hotel

A ladybird peeks from her cosy hiding place in a bug hotel.

Ladybird

The little **beetle** is hungry.

It is time to go hunting!

The ladybird lifts up her red, spotty wing cases.

Wing case

Wing

She unfolds her long, see-through wings.

Whoosh

Off she flies to look for food.

Ladybirds and other beetles are all **insects**.

The ladybird lands on a plant.

There are tiny insects called aphids on the plant's leaves.

White aphid

Munch Munch Munch

Aphids are a ladybird's favourite food.

Black aphid

A ladybird eats about 50 aphids each day.

Sometimes the ladybird eats **pollen** from flowers.

Pollen

When she needs a drink,
she sips water from raindrops.

A hungry bird sees the ladybird. Is she in danger?

No! The ladybird's red colour tells the bird that she tastes bad.

Ladybirds make a stinky, yellow gloop that comes out of their legs.

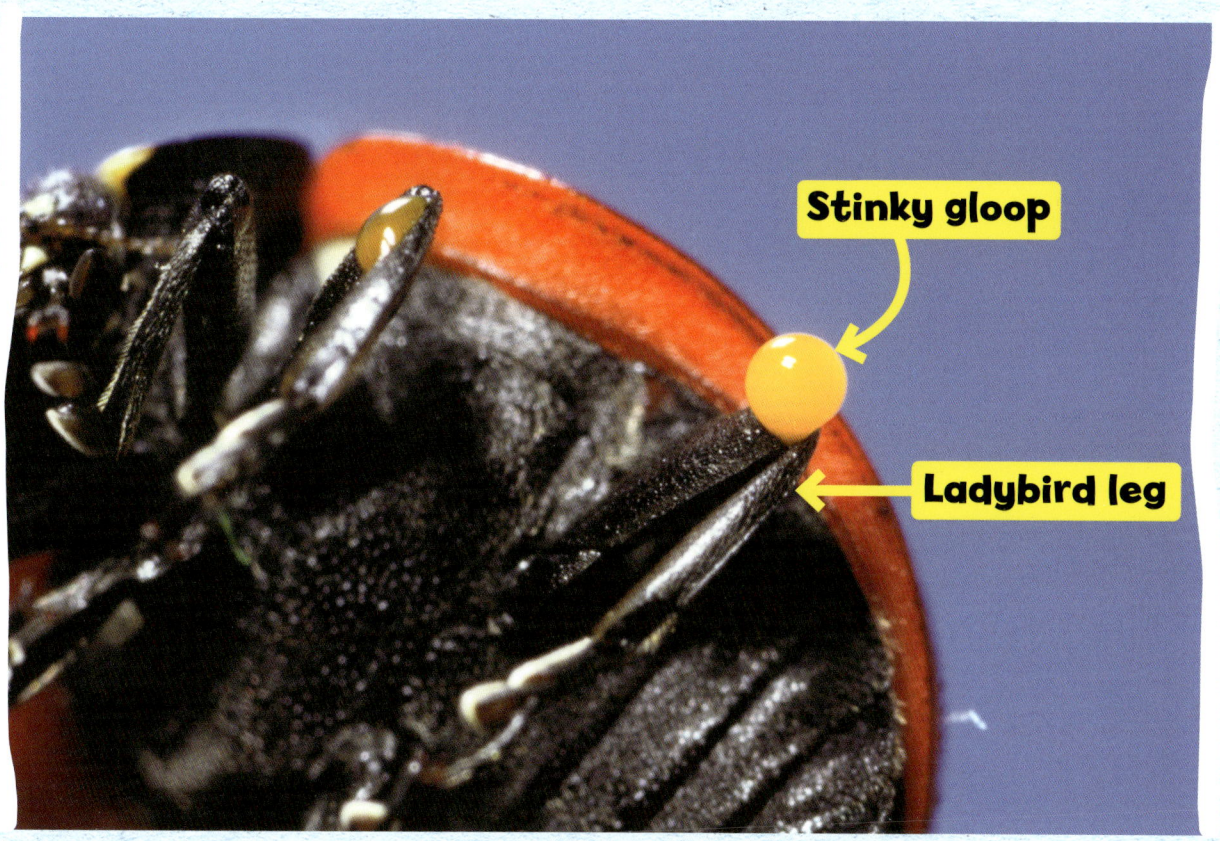

Stinky gloop

Ladybird leg

This stops **predators** eating them.

A few days ago, the ladybird **mated** with a male ladybird.

Today, she is ready to lay eggs.

The ladybird lays her tiny yellow eggs under a leaf.

Ladybird eggs

A ladybird can lay about 300 eggs.

After laying her eggs, the ladybird eats more aphids.

Aphids

Munch Munch Munch

Then she finds a cosy hiding place to spend the night.

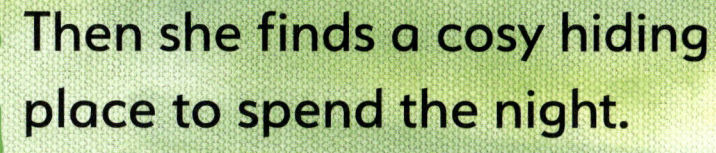
Tree bud

What will happen to the ladybird's eggs?

A baby ladybird called a **larva** hatches from each egg.

Munch Munch Munch

The larva eats aphids and gets bigger.

Pupa

After three weeks, the larva becomes a **pupa**.

Now lots of changes happen.

After a few days, the pupa's case splits open.

A tiny yellow beetle climbs from the case.

After a few hours, the insect turns red and grows spots.

It's a new ladybird!

Glossary

beetle
A type of insect with wings and two hard wing cases or covers.

Wing case
Wing

insect
A tiny animal with six legs. Ladybirds, butterflies and bees are all insects.

larva
A young insect that hatches from an egg.

Ladybird larva

mate
To come together to produce young.

pollen
Dust made by flowers that helps plants make seeds and new plants.

predator
An animal that eats other animals.

pupa
A stage in the life cycle of some insects. Inside its pupa case, an insect changes into an adult.

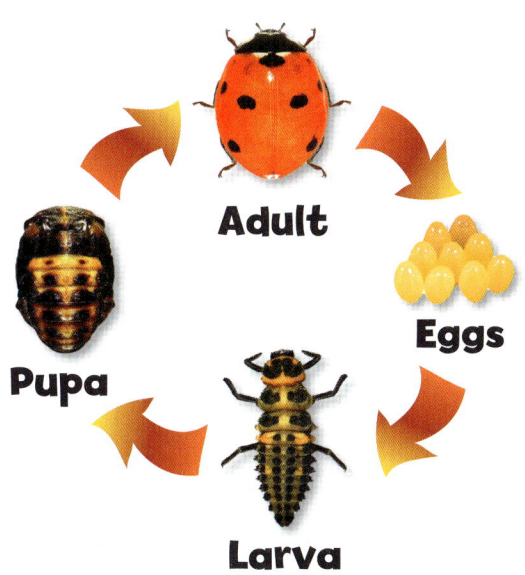

Index

A
aphids 8–9, 16, 18

E
eggs 14–15, 16–17, 18

F
food 5, 7, 8–9, 10, 16, 18

L
larvae 18–19

M
mating 14

P
pollen 10
predators 12–13
pupae 19, 20–21

W
wings 6–7